539.7
ENG

Engdahl, Sylvia Louise

The subnuclear zoo.

008634

BOOKS BY SYLVIA ENGDAHL

Enchantress from the Stars
Journey Between Worlds
The Far Side of Evil
This Star Shall Abide
Beyond the Tomorrow Mountains
The Planet-Girded Suns

Anywhere, Anywhen: Stories of Tomorrow
edited by Sylvia Engdahl

Universe Ahead: Stories of the Future
*Selected and Introduced by
Sylvia Engdahl and Rick Roberson*

the sub-
nuclear zoo

New Discoveries
in High Energy Physics

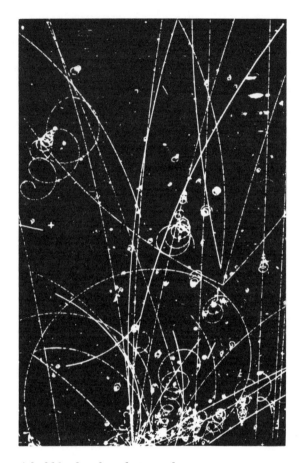

A bubble chamber photograph.

the sub-
nuclear zoo

New Discoveries in High Energy Physics

by SYLVIA ENGDAHL and RICK ROBERSON

ILLUSTRATED WITH DIAGRAMS,
CHARTS AND PHOTOGRAPHS

MEDIA CENTER
CASTILLERO JUNIOR HIGH
SAN JOSE, CALIFORNIA

Atheneum   New York   1977

LIBRARY OF CONGRESS CATALOGING IN PUBLICATION DATA
Engdahl, Sylvia Louise. The subnuclear zoo.
   Includes index.
   SUMMARY: An introduction to high energy physics, including a description of sub-atomic particles and a discussion of current theories in the field and areas for future research.
   1. Particles (Nuclear physics)—Juvenile literature.
[1. Nuclear physics]   I. Roberson, Rick, joint author.
II. Title.
QC793.27.E53   539.7   77-1686
ISBN 0-689-30582-6

Copyright © 1977 by Sylvia Engdahl and Rick Roberson
All rights reserved
Published simultaneously in Canada by
McClelland & Stewart, Ltd.
Manufactured in the United States of America by
The Murray Printing Company
Forge Village, Massachusetts
Designed by Mary M. Ahern
First Edition

Contents

THE INNER MAKEUP OF MATTER   3

PARTICLES OUTSIDE ATOMS   21

WAYS OF OBSERVING THE INVISIBLE   40

THE MEMBERS OF THE ZOO   61

SOME MYSTERIES STILL TO BE SOLVED   81

Index   99

the sub-nuclear zoo

New Discoveries in High Energy Physics

The Inner Makeup of Matter

What are things made of?

Since the beginning of human history, people curious about the world have wondered what things are made of. They have wanted to know what things are like inside, and what makes one substance different from another. Why, for instance, is a piece of metal different from a rock? Anybody can see that it *is* different, but what causes the difference? What makes water unlike earth, and the planet Earth itself unlike the stars? These questions have been asked for thousands of years.

Scientists have learned the answers to many such questions. At times in the past, it has seemed that their explanations could account for all that is known about what matter is like, and how various kinds can be used. But these answers have never been complete answers. The more scientists learn in answering old questions, the more new questions they find to ask. So, although scientists today know a great deal about the structure of matter, they are still searching for a simpler way to explain what things are made of.

One of the goals of science is to find simple, basic explanations that tie many complicated facts together. For instance, long ago people wondered what held the planets in the sky and what made them move along such exact paths. Various explanations were suggested, some of which were believed by scholars for centuries. But as more was observed about planets, those explanations had to be made more and more complicated to fit the observations. Because of this, scientists began to be dissatisfied with their theories. Then, in the seventeenth century, a new theory was developed that introduced the idea of *gravitational force*. Formulating mathematical laws of gravitational force was extremely complicated, but the laws themselves provided a much simpler way to explain the movement of planets—and also other things—than had existed before.

For a long time people believed that the laws of gravitation were a final answer to many questions about the universe. But those laws concerned how things move, not what they are made of. There did not seem to be much connection between the two problems. Meanwhile, scientists kept on looking for answers to their questions about the structure of matter. Eventually, they did see a simple pattern that explained the complex facts about the differences between various substances. And as usual, this answer led to new questions, and to discoveries that did not fit what was believed.

These discoveries involved some surprises. One surprise was that the question of what things are made of cannot be separated from the question of how things move. All matter is in motion. The movement of its invisible components is related to what holds it together. Gravitational laws were found to be only part of a big pattern; they did not explain all motion as well as they explained the orbits of planets. There are forces besides gravitation that people of the seventeenth century did not know about. Scientists had to admit that their theory was not as complete and final as it had once seemed.

Today, most scientists are more cautious about believing they see complete patterns than were scientists of the past. Scientific discovery progresses faster than it used to, because new knowledge and new methods of experimenting have caused new questions to appear at a more rapid pace. Surprises occur rapidly now, too. Men and women working in science expect to be surprised; that is what makes the work exciting. One of the most exciting areas of science, at present, is the search for further knowledge about what things are made of.

In the seventeenth century, and even much later,

The Subnuclear Zoo

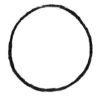

Nobody knows just what shape particles have. They are usually drawn as if they were round because that is the simplest way to show them.

scientists studied how things are structured by looking at objects large enough to see. But the questions that can be answered that way have already been answered. Now people are asking about the *particles* of which matter is composed, which are too small to see even with microscopes. There are many different varieties of particles, so many that scientists sometimes refer to them as "the subnuclear zoo." The study of such particles is called *particle physics*. It is also called *high energy physics* because of the kind of experiments needed to find out about particles without seeing them.

This book tells some of the most important facts physicists have learned about the structure of matter. It also tells about questions that are being studied now, in the year of America's bicentennial. By the time you read this book, some of these questions may have been answered and still newer ones may have been raised. Probably no book will ever give a complete, final explanation of what the universe is made of. All the same, reading about what is now known, and what people now wonder, will remain a way of discovering the right questions to ask.

Things are made of atoms

By the late nineteenth century, scientists thought they knew the whole story of how matter is constructed: things are made of *atoms*. An atom is the smallest bit of a particular kind of matter that can exist. The smallest bit of iron that can exist is an iron atom, the smallest bit of silver that can exist is a silver atom, and so forth. There are about a hundred different kinds of atoms, and these kinds are called *elements*. Iron is one element and silver is another. Obviously there are a great many more substances on Earth than a hundred, but most of them are chemical combinations, or *compounds,* of two or more elements. For example, water is a compound of the elements hydrogen and oxygen. This means it has both hydrogen atoms and oxygen atoms in it. Nineteenth-century scientists studied "what things are made of" in terms of how atoms combine to form compounds.

The knowledge of chemistry gained through this study was, and still is, important knowledge; it is what makes most of our technology possible. And, in itself, it was correct. Things are made of atoms and nothing else. The same kinds of atoms exist all over the universe. The same elements are found in the stars as on

This galaxy is a group of stars very far from Earth, but the atoms of the elements there are exactly like the atoms of the same elements here.

The Inner Makeup of Matter

Earth. Hydrogen atoms in distant galaxies are exactly like the hydrogen atoms in the water of Earth's oceans.

Everything in the universe is made of atoms. But what are atoms themselves made of?

An atom is very small, much smaller than is easy to imagine. A hydrogen atom is only about a hundred millionth of a centimeter in diameter. There are about one thousand billion billion atoms in a large drop of water. It's quite understandable that scientists once believed that there could be nothing smaller than an atom.

But if there were nothing smaller than an atom, there would be no way to account for the fact that there are different kinds of atoms. The atoms of different elements are different. What is different about them? How is an oxygen atom unlike a hydrogen atom? When scientists began to ask these questions, they realized that atoms could not be solid balls of matter, as was first thought. They had to have some sort of internal structure that made them unlike each other. In other words, the atoms themselves had to be composed of particles.

Although the atoms of one element are not like those of another element, the particles of which those atoms are composed are identical. Particles in an oxygen atom are exactly like those of a silver atom, or those in an atom of iron. What makes the atoms different is the *number* of particles in each kind. There are many types of particles—no one yet knows how many—but only one

Each element has a different number of protons in its atoms. This number is called its atomic number.

○ HYDROGEN, Atomic Number 1

○○ HELIUM, Atomic Number 2

○○○○○○ CARBON, Atomic Number 6

○○○○○○○○ OXYGEN, Atomic Number 8

type determines the difference between the elements. This is a particle called a *proton*. Counting protons was the simple pattern that explained each element's uniqueness. An atom of hydrogen has one proton in it. An atom of oxygen has eight protons. An atom of iron has twenty-six. The heaviest natural element found on Earth, uranium, has ninety-two protons in each atom. Scientists call the number of protons an element has in each atom its *atomic number*. They have found an element for every number from 1 to 92, plus a few elements with higher atomic numbers that can be made in laboratories.

Protons are important because the number of protons in an element's atoms is what distinguishes that element from other elements. But atoms are not just clusters of protons. When scientists first began to speculate about the particles in atoms, they knew that there must be more than one type. In fact, before they found protons, they discovered *electrons*. They made this discovery because they were studying electricity.

The Inner Makeup of Matter

Atoms contain electricity

Electricity is not something present only in wires that carry electric current. It is everywhere: in the sky, in the ground, in human bodies, in everything we see and touch. Most people picture electricity as something that has to be generated to produce power. It is true that electric current must be generated for electric power to be harnessed and used. Electricity itself, however, is part of every substance. Scientists did not know this until the end of the nineteenth century.

They had known something about electricity long before then. They had known, for instance, that there are two kinds called *positive* and *negative*. The amounts of positive electricity and negative electricity in a substance are normally equal. When they are equal, the electricity cannot produce any noticeable effects. However, a state can occur in which there is more positive electricity than negative, or vice versa. When an object is in such a state, it is said to have an electric *charge* or

There are two kinds of electricity, positive and negative. A plus sign means a positive electric charge. A minus sign means a negative electric charge.

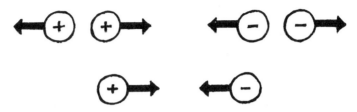

Electric charges of the same kind repel each other. Electric charges of opposite kinds attract each other.

to be charged. A charge, which is either positive or negative, is a case of one kind of electricity exceeding the other kind.

Electrical charges of the same kind repel each other, or push each other away. Electrical charges of opposite kinds attract each other, or pull toward each other. If two things with positive charges—or two things with negative charges—are near each other, they will move farther apart unless they are affected by a force stronger than the electrical repulsion between them. If, on the other hand, something with a positive charge is near something with a negative charge, they will draw closer together unless affected by a force stronger than the electrical attraction.

Nineteenth century scientists knew these facts about charges, and they knew ways of charging things. They knew ways of harnessing electricity, too. But physicists did not really understand what was happening until

The Inner Makeup of Matter

they began to find that atoms are made of particles. Then they began to realize that the question of what things are made of is closely related to the question of how things move, for particles are in constant motion.

The first thing they learned about particles was that small particles with negative electric charges can be separated from atoms, and they named these particles electrons. But they were aware that if there are electrons in atoms, there also have to be particles with positive electric charges. Otherwise everything would have a negative electric charge all the time. For an object to be in its normal uncharged state, the positive and negative electricity in it have to be equal. So physicists believed that protons, which are positively charged particles, exist. However, the protons were not as easy to find as the electrons because of the way atoms are structured.

Atoms are not solid

It seems as if the atoms of solid elements must be very solid indeed, without any space in them. A table, for instance, seems solid. A rock or an iron bar seems still more hard and firm. But actually, even the heaviest elements are not solid at all. They are made of atoms, and atoms have more empty space in them than space filled with particles.

When scientists first learned about atomic structure,

PROTON   ELECTRON   NEUTRON

The three particles that form atoms are protons, electrons and neutrons. Protons have a positive electric charge. Electrons have a negative electric charge. Neutrons have no electric charge at all.

they drew diagrams of atoms that looked like little solar systems. They believed an atom really was like a miniature solar system, with the electrons orbiting around a solid ball of protons. Such diagrams are still useful for picturing an atom. Today, however, scientists know that the analogy between the shape of an atom and the shape of a solar system is not perfect. The electrons do not have precise orbits that can be calculated, as orbits of planets around the sun can be calculated. According to present theory, their exact paths cannot ever be determined.

Nevertheless, the electrons do circle the center of the atom, and they are a long way from the center in comparison to its size. The center of an atom is called the *nucleus*, and all the particles except the electrons are inside it. If the nucleus of an iron atom were as big as a baseball, the outermost electrons would be spinning around it more than a kilometer away. By far the largest part of an atom is the space between the nucleus and the electrons.

The Inner Makeup of Matter

Nearly all of the weight of the atom is in the nucleus, however. The protons, for example, are almost two thousand times heavier than the electrons are. There are also other heavy particles called *neutrons* in the nucleus, which have an important role in holding it together.

The electrons are not as firmly "attached" to the atom as the particles in the nucleus. They are held in their orbits by electrical attraction or—in more exact terms— by *electromagnetic force*. Because electrons have negative electric charges, they are attracted to the protons, which have positive electric charges. Opposite charges attract each other; therefore each proton in the nucleus attracts an electron, and as many electrons as protons are

This is a way of imagining how an atom might look if it could be seen. The electrons are really farther from the nucleus than a picture can show, and they do not stay in exact paths. Today, scientists use diagrams that tell more about how particles act than how they look.

normally in an atom. A hydrogen atom, with one proton, has one electron held in orbit by electrical attraction. An iron atom, with twenty-six protons, has twenty-six electrons in orbit. The neutrons have no electric charge, so they have no effect on the number of electrons in the atom.

Although electromagnetic force keeps the electrons in orbit, they do not stay there permanently. There are many conditions in which some of the electrons can leave their orbits. For instance, when an electric current passes through a wire, electrons in one atom of the wire move to another atom. That is what an electric current is: the movement of electrons from atom to atom in a specific direction. Atoms do not change very much if some of their electrons are temporarily lost. It is the protons, not the electrons, that determine the nature of an atom. The atom is still the same element as long as the number of protons does not change.

What keeps the protons in the nucleus? After all, they all have positive electric charges, and objects with the same kind of electric charge repel each other. If there were no force stronger than electromagnetic force, the protons would not cling together; they would push each other apart. Scientists wondered why protons don't do this, and that is when they began to learn about *nuclear force.*

The Inner Makeup of Matter

Nuclear force holds the nucleus together

To many people, the term "nuclear force" means something to do with bombs or power plants. Those are the subjects most often being discussed when this term is used in newspapers or on television. Such use has given people the misleading impression that nuclear force is something unnatural that should not, perhaps, have been "invented."

Actually, nobody "invented" nuclear force. It is one of the most fundamental forces in nature. The word *nuclear* simply means having to do with the nucleus. Nuclear force is what holds the protons and neutrons together in the nucleus of an atom. If it did not do this, there would be no atoms—no elements. The earth and the sun and everything else in the universe would fall apart. Nuclear force is a very basic part of the answer to questions about what things are made of.

Nuclear reactions—processes involving changes in the nuclei of atoms—are important in nature, too. The heat and light of the sun are produced by a nuclear reaction taking place within it: the fusion of hydrogen atoms into helium atoms. Nuclear fusion may someday become a safe and unlimited source of power on Earth. Today's nuclear power plants involve fission—the splitting of the nucleus in uranium or plutonium atoms—rather than

The Subnuclear Zoo

fusion. Using fusion for generating power will require a big scientific advance. Just as scientists have learned to harness electromagnetic force to do useful work with electricity, they are learning to harness nuclear force. Nuclear energy is terrible when it is used for weapons, but in itself, it is as natural and essential as sunlight.

When people speak of nuclear energy, they generally mean the use or release of that energy. The energy is already in everything, bound up in atoms, holding them together: in water, or in the human body, just as it is in the sun. A table or a chair contains the same kind of nuclear energy as exists in a nuclear bomb. What happens in the bomb is that some of this energy is released in the form of an explosion. Most releases of nuclear energy are not explosions, and only under very extraordinary conditions can any releases at all be produced.

Nuclear force is involved in any process that affects the nucleus of an atom. Though bombs are a poor example of such a process, thinking of them does give an idea of how much energy holds the nuclei of atoms together. It is fortunate that the force is so strong since that is what makes matter stable and permanent. Yet the very strength of nuclear force makes the particles in the nucleus of an atom hard to study. Separating them enough to study them requires the use of electrons or protons that have been given extremely high energies. That is why this study is called high energy physics.

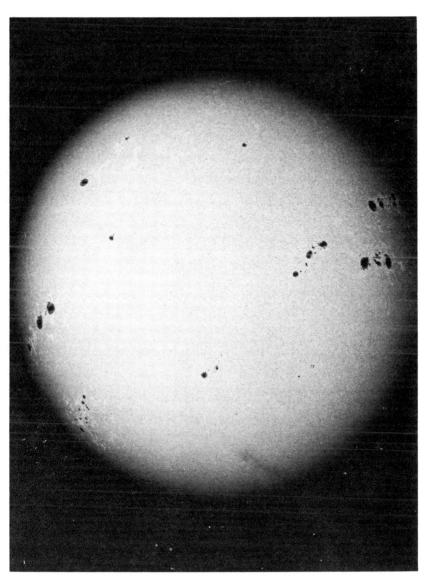
Nuclear fusion of hydrogen atoms into helium atoms is what makes the sun shine.

The Subnuclear Zoo

Of course, most experiments in high energy physics do not make use of nuclear fission or fusion reactions, as the generation of nuclear power does. And all high energy physics experiments take place on a small scale. Scientists do not need to study large numbers of atoms; what they want is to find out about the particles in representative atoms.

They already know that these particles are alike no matter what kind of atom they come from. This concept is immensely significant. To say that things are made of identical particles is more basic than to say that things are made of different atoms. It is a step toward the simple answer scientists are looking for.

But only a step. The search for such an answer is very complicated, because nuclear particles are sometimes transformed into *subnuclear* particles. Whether they are originally made of subnuclear particles or whether all the subnuclear particles come into existence during the transformation is something physicists are not yet sure about. However, subnuclear particles can be observed during experiments. These particles make up the subnuclear zoo, and there are more of them than anyone imagined at first. By the time this book is read, scientists will probably be speculating about subnuclear particles they had not imagined at the time it was written.

Particles Outside Atoms

Not all particles stay in atoms

All elements are made of atoms. Atoms are made of particles called electrons, protons and neutrons. From protons and neutrons come the subnuclear particles that scientists are learning about. But if particles always stayed inside atoms, it would be difficult to learn much about them. Atoms are too small to see except by means of special microscopes, and the protons and neutrons in their nuclei are even smaller. Furthermore, particles interact with each other, so that when they are bound together, they do not behave in the same way as they do separately.

The nucleus of an atom is held together by nuclear force. In most cases the composition of the nucleus cannot change, although there are many conditions under which the number and location of the electrons sur-

THE SUBNUCLEAR ZOO

rounding it can change. Because electrons do move from atom to atom, scientists discovered them before they discovered protons and neutrons. The nuclear particles could not be effectively studied inside atoms, and it was not known that nuclei ever change.

An atom with an unchanging nucleus is called *stable*. Fortunately, most atoms on Earth are stable; that is why the elements made of those atoms are permanent. However, there are cases where the nuclei of atoms are not stable. This is fortunate, too; because if it were not true, early atomic scientists would have found it hard to study the structure of matter.

Certain elements are naturally unstable. These elements are not very common, and they are not in things most people possess, but scientists have found important uses for them in addition to studying them. Other elements can be made unstable in laboratories, although

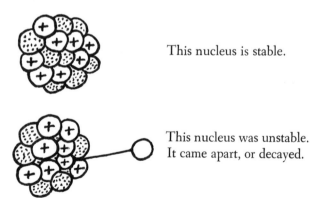

This nucleus is stable.

This nucleus was unstable. It came apart, or decayed.

they do not become unstable by themselves on the planet Earth.

When an element is not stable, the nuclei in the atoms of that element can come apart. Some of the protons and neutrons can leave the atom, and when they do, they sometimes turn into other particles. They can also give off other particles without leaving the atom. These processes are called *decay*. Since subnuclear particles can be studied only outside atoms, particle physics is largely the study of decay. There are many kinds of decay that particle physicists are studying.

Particles outside atoms are called radiation

Radiation is a familiar word. The sun gives off radiation; people see rays of sunlight and feel their warmth. A light bulb gives off radiation. So does a hot stove. All sources of light and heat emit radiation, and there are many additional kinds of radiation that cannot be seen or felt. But what is radiation made of?

Whenever particles exist apart from atoms, they are called *radiation* by scientists. Not all the particles of which radiation consists are particles of matter. Scientists use the word "particle" to refer to bits of pure energy as well as bits of matter. They do this because matter can be transformed to energy, and vice versa. It was once thought that matter and energy were completely unrelated and that the universe would always have the same amounts of each that it had to begin with. However, through nuclear physics, scientists have learned that matter and energy are different forms of the same thing. Now, only the total amount of both is believed to remain constant. Energy particles must therefore be considered when the structure of matter is studied.

The different types of radiation are classified by the particles of which they are composed. Radiation is also often classified by the kind of process that produces it or by the amount of energy its particles have. No simple list of radiation types can be made because the various classifications overlap.

All radiation carries energy, and sometimes matter, from one place to another. It does this because it is composed of particles, and all particles move. Particles that are pure energy move at the speed of light, which is as fast as anything can move according to the theories scientists now believe. This speed is called "the speed of light" because light was the first form of pure energy scientists measured, but it might be easier to think of it simply as "the speed of energy." Particles that have *mass* —the term physicists use for amount of matter—in addition to energy cannot move at the speed of light, although sometimes they can move almost that fast.

Scientists believe that all forces work by means of

energy-carrying particles. They know this is true of electromagnetic force and one kind of nuclear force. They are not quite sure about gravitational force because they have not been able to prove the existence of any particles that carry it.

Electromagnetic force is carried by *electromagnetic radiation*. Electromagnetic radiation is composed of particles called *photons*. Photons are pure energy; they do not have any mass. Naturally, they move at the speed of light. Light itself is simply a kind of electromagnetic radiation. It is not the only kind—radio waves are electromagnetic radiation, too, for example. Electromagnetic radiation is commonly discussed in terms of waves instead of in terms of particles. Not only is this the most convenient way to talk about it, but scientists sometimes find the idea of a wave is the best description of it. The mathematics used for analyzing behavior of particles often requires them to be considered as waves. Photons do not move in wavy paths, however. Strange as it may seem, the difference between a particle and a wave is something scientists themselves do not fully understand.

Since electromagnetic force is what makes particles with negative electric charges and those with positive electric charges interact, it seems as if electromagnetic radiation might contain electricity; but that is not the case. Electromagnetic radiation is composed entirely of photons, and photons do not have any electric charge.

The Subnuclear Zoo

More energy

> GAMMA RAYS
>
> X RAYS
>
> ULTRAVIOLET LIGHT
>
> VISIBLE LIGHT
>
> INFRARED LIGHT
>
> MICROWAVES, RADAR
>
> TV, FM RADIO
>
> SHORT WAVE RADIO
>
> AM RADIO

Less energy

Photons make up electromagnetic radiation. They can have different amounts of energy. Photons that make up visible light have more energy than those that make up radio waves, but less energy than those that make up X rays.

They carry the force that makes electromagnetic interactions occur, but the electricity is already in the particles of matter between which this force is carried.

Where does electromagnetic radiation come from? It is given off, or emitted, by atoms. This radiation may travel long distances across space. The photons from a light bulb come from atoms inside the bulb; the photons that make up the light from a star come all the way from the atoms in the star. Of course, not all photons are visible as light, since light is not the only form of electromagnetic radiation. The form of electromagnetic radiation depends on the amount of energy the photons have when they leave the atoms, and the amount of energy they have depends on what type of process causes them to be emitted.

Although photons are given off by atoms, they are not particles that make up atoms like electrons, protons and neutrons. They originate during the process that gives them off. Sometimes it is a process involving the atom's electrons only. At other times it is a nuclear decay process. Certain nuclear decay processes do give off electromagnetic radiation. However, most of them emit other types of radiation, which means particles other than photons. Physicists are not yet sure whether all such particles exist in some form within protons and neutrons or whether they too originate during the decay, or transformation, process.

Radioactivity is a nuclear decay process

Elements that are unstable are called *radioactive* elements. Almost everyone has heard that radioactive elements give off dangerous radiation. It can indeed be dangerous when untrained people come into contact with it. Scientists know what precautions to use, however, and they have special equipment that enables them to study and use radioactive materials safely.

By studying naturally radioactive elements, scientists first learned that there are nuclear particles. When such particles are emitted from atoms in the form of radiation, it is possible to discover what the particles are like in themselves. There are several types of nuclear radiation, depending on the kinds of particles of which the "rays" are composed. It is important to remember that these "rays" are not necessarily streams of particles; physicists often use the word "ray" when they are talking about a single particle or small group of associated particles. They began to do this before they understood what nuclear radiation is like.

What makes some elements radioactive when the rest are naturally stable? This depends on the number of neutrons in the element's atoms. The number of protons in an atom determines which element it is, but the number of neutrons determines whether or not the element is stable. The neutrons play a part in holding the nucleus of an atom together through nuclear force. Nuclei with too many or too few neutrons eventually decay.

Although all atoms of the same element have the same

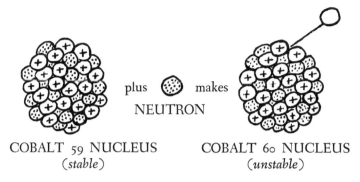

Scientists can add a neutron to the nucleus of a cobalt 59 atom and make a radioactive cobalt 60 atom.

number of protons, they do not necessarily have the same number of neutrons. Naturally stable elements can be made radioactive in laboratories by changing the number of neutrons in their atoms. Forms of an element with different numbers of neutrons are called *isotopes*. It is possible for unstable, or radioactive, isotopes of an element to be made even if the isotopes of that element found on Earth are stable. For example, the element cobalt has 27 protons and its natural isotope has 32 neutrons, which gives it a total of 59 in its nucleus. The isotopes are named by this total number, so normal cobalt, which is stable, is called cobalt 59. An artificial isotope of cobalt with 33 neutrons is called cobalt 60; this is a radioactive isotope used in the treatment of cancer.

Not all of the atoms of a radioactive element decay at

the same time. Scientists do not know what makes an atom decay when it does, rather than at some other time, so they cannot predict when a particular atom is going to decay. However, they can predict how long it will take for half the atoms in a radioactive element to decay even though they cannot tell which ones. They can do this because no matter how much or how little of a given element they study, it always takes the same length of time for the decay of half its atoms to occur. This length of time is called the element's *half-life*. It may seem as if the second half of the element would decay in the same length of time that the first half did, but that is not what happens. Only half of the remaining portion decays. The same proportion of atoms, not the same number of atoms, decays during each half-life period.

Although the instability of radioactive elements is a result of there being too many or too few neutrons in proportion to protons, the number of neutrons is not the only thing that can change during radioactive decay. The number of protons often changes, too. This means radioactive decay can transform one element into another element. All radioactive elements gradually turn into stable elements through natural decay processes. How long this takes depends on the half-lives of the various radioactive isotopes created during the decay sequence. Half a sample of the artificial isotope copper

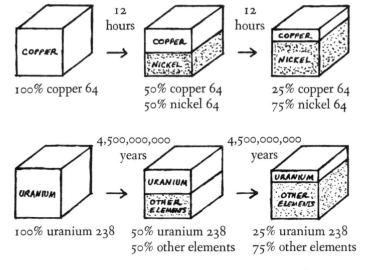

Different radioactive elements decay at different rates. Only half the atoms of the element decay in each half-life period. The atoms that decay are scattered throughout the material, although here they are drawn as if they were all at the bottom. Some elements go through many separate decays before a stable element is produced. Uranium is one of these; there is not room here to list all the steps by which it turns into lead.

64 decays into the stable isotope nickel 64 in about twelve hours. It takes millions of years, however, for half of a uranium 238 sample to turn into stable lead 206.

There are three kinds of radioactive decay

There are three kinds of natural radioactive decay, and the particles that make up the radiation are different in each kind. These kinds are named *alpha decay, beta decay* and *gamma decay*. The radiation is called alpha radiation, beta radiation and gamma radiation, so the particles of which it consists are sometimes called alpha, beta and gamma particles. However, all these particles also have other names. They are not always produced by radioactive decay, and they sometimes exist without being part of nuclear radiation.

The first type of radioactive decay scientists discovered was alpha decay. In alpha decay, two protons and two neutrons leave the nucleus of a radioactive atom. Therefore, an alpha particle is made of two protons and two neutrons. This is just the same as the nucleus of a helium atom. Scientists say that alpha particles are helium nuclei, but they do not call helium nuclei alpha particles. They speak of alpha particles

ALPHA PARTICLE
(HELIUM NUCLEUS)

During alpha decay, a radioactive nucleus emits an alpha particle, which is made of two protons and two neutrons. An alpha particle is the same as a helium nucleus except for having more energy.

only when these particles are products of radioactive decay.

The alpha particles that make up alpha radiation are not harmful in themselves; after all, they are just protons and neutrons. They are dangerous when they exist as radiation only because they are moving at high speeds. They can cause damage by striking at high speed just as a bullet, which is safe to touch, causes damage when it is shot from a gun. Alpha particles are the least dangerous type of nuclear radiation because they are less penetrating than the particles in the other types. They can be stopped by thin materials. The dial of a luminous watch gives off alpha radiation because a naturally radioactive element, radium, is used in the paint, but the alpha particles do not go through the glass face of the watch.

In alpha decay, because protons leave the nucleus of the atom, the atom is transformed into a different element that has an atomic number two less than the radioactive element did. If that element is also radioactive, more decay occurs until the element has been transformed into a stable one. This is not just a matter of subtracting protons two at a time, however, because neutrons are emitted too. The change in the proportion of neutrons left in the atom often causes other types of nuclear decay to occur along with alpha decay.

Alpha radiation is not made of subnuclear particles.

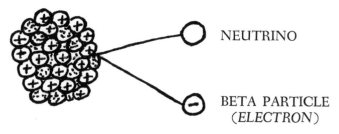

During beta decay of naturally radioactive elements, a neutron in the nucleus decays into a proton, electron and neutrino. The proton stays in the nucleus. Both the electron and the neutrino are emitted, but only the electron is called a beta particle.

It is simply nuclear particles that leave the atom. Beta decay is more complicated. Through studying beta decay, scientists first learned that other particles are emitted by nuclei. They also learned that there is more than one kind of nuclear force.

In beta decay of naturally radioactive elements, a neutron within an atom of the element undergoes a transformation. It turns into three particles: a proton, an electron, and another particle called a *neutrino*. The proton stays in the nucleus of the atom, which means that the atom is changed into a different element with an atomic number one higher than the original element. For example, phosphorus 32, which has 15 protons and 17 neutrons, is changed into sulfur 32, with 16 of each.

When a neutron is transformed by beta decay, the electron that is created flies away from the atom at high

Particles Outside Atoms

speed. It does not stay in the outer part of the atom with the other electrons because it has too much energy to be held by the electromagnetic attraction of the atom's protons.

Scientists can calculate in advance how much energy will be released by the transformation of a particle. They knew how to do this even before they had learned that there are subnuclear particles observable only as radiation. They also knew how to measure the energy of the particles actually emitted and the energy left in the atom. When they studied beta decay, they found that these measurements did not match their calculations. The advance calculations showed that there should be more energy than was measured in their experiments. This told scientists that there must be some particle besides the electron carrying away energy. They named it "neutrino," which means "little neutral one," because they knew it had to have a neutral electric charge and that it could not be very heavy compared to a neutron. In fact, it could not be a particle of matter at all, because if it were, the mass of the atom would be reduced by its emission, and this was not the case. A neutrino, like a photon, is a particle of pure energy. Scientists did not actually find neutrinos until long after they knew such particles must exist.

There was something else that puzzled scientists when they began to study beta decay. They did not understand what force caused the neutron to be transformed. The only forces known to exist were gravitational force, electromagnetic force, and nuclear force. The first two do not cause significant effects within single nuclei. Nuclear force does, but no explanation based on it could account for the observed facts.

Beta decay showed scientists that there is more than one kind of nuclear force. They named the kind responsible for beta decay the *weak nuclear force* or *weak interaction*. That is not a convenient name, because it makes it necessary to call ordinary nuclear force the *strong nuclear force* to keep the two kinds straight, which in some ways is misleading. Sometimes strong nuclear force is called *nuclear binding force*, but usually physicists stick to calling it the strong force.

Weak nuclear force takes longer to act than strong nuclear force. Also, it has a very short range—it works on a scale too small for the strong nuclear force to have any effect. Particles must be extremely close together for a comparatively long period to interact by weak nuclear force. Because of these facts, it is "weak" in the sense that the chances of its causing interactions between particles are small. Nevertheless, weak interactions among subnuclear particles do occur, and scientists know that the weak force is what causes beta decay, though they do not yet fully understand how the force works.

Although two particles—an electron and a neutrino

Particles Outside Atoms

—leave the atom during deta decay, only the electron is called a beta particle. This is another case where a name was given before everything about the process was discovered. Many such names are used in particle physics. Scientists have to name particles and their characteristics in order to talk about them, and later, when the names are found to be confusing, it is too late to change.

Beta radiation is more dangerous than alpha radiation because beta particles are more penetrating than alpha particles, and more shielding is therefore needed to stop them. However, beta radiation is very useful when it is harnessed. For instance, doctors can use beta radiation to treat some types of cancer.

Gamma radiation is also used to treat cancer. Gamma decay does not change the number of protons in an atom and, therefore, does not change one element into another. It does not even change one isotope of an element into another isotope, for it causes no change in the number of neutrons, either. It merely releases extra energy from the nucleus that is left over following some

GAMMA PARTICLE (PHOTON)

During gamma decay, a radioactive nucleus emits a gamma particle, which is a photon with very high energy.

previous nuclear reaction. This energy is emitted in the form of gamma particles, which are simply photons. The photons of which gamma radiation is composed have much more energy than most other photons, however, and it is the most dangerous form of nuclear radiation. It requires more careful handling than the other types, but can do great good when controlled.

Subnuclear particles also decay

Radioactive decay is the only kind of nuclear decay that occurs spontaneously on the planet Earth. It is not the only kind of decay that exists, however. All subnuclear particles except photons and neutrinos are unstable, which means that they decay into stable particles. In other words, they turn into protons or electrons, sometimes giving off photons or neutrinos in the process.

Subnuclear particles decay only outside atoms in the form of radiation. Scientists did not begin to learn about them until they understood what radiation is. Then they discovered that there are unstable particles, or "rays," as well as unstable atoms. These particles have very short lifetimes; they decay in mere fractions of seconds. Often there are several steps to the decay sequence: one subnuclear particle can turn into other subnuclear particles, which continue to decay into still others. In the end, stable particles are produced.

Particles Outside Atoms

Unstable particles have such short lives that they must be created just before they are detected. Otherwise, they would not last long enough for anyone to find out about them. Scientists have learned that these particles come into being only when something causes stable particles to be transformed. Usually this is a result of smashing the nuclei of atoms, which is why the particles are called subnuclear. Subnuclear particles should not be pictured as fragments of smashed nuclei, however. In fact, some have more mass than the particles of which atomic nuclei are composed. The question of whether any subnuclear particles exist all the time inside nuclear particles is complex, and physicists themselves do not yet have a complete answer.

Physicists do know, however, where free subnuclear particles get "extra" mass. They get it from energy—mass and energy are, after all, two forms of the same thing. Stable particles do not combine to form free subnuclear particles unless they collide at very high energies. This is why high energy experiments are needed to study the "subnuclear zoo."

Ways of Observing the Invisible

High-energy particles can come from space

Subnuclear particles produced by processes other than radioactivity do not come into being spontaneously and are not normally found on Earth. They are given independent existence when the nuclei of atoms are smashed by collisions with each other or with electrons traveling at high speeds. Collisions that smash nuclei, or transform particles within nuclei, occur only when at least one of the protons or electrons involved has extraordinary energy. Just as a collision between automobiles does not smash them if they are both moving slowly, a collision between particles does not transform them if they are both moving at relatively low speeds.

There are no natural forces on Earth that produce particles with enough energy to smash the nuclei of atoms. However, there are many such particles in

The Subnuclear Zoo

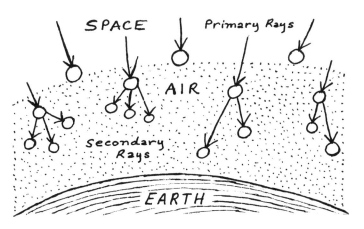

Primary cosmic rays come from space. When they collide with the air, secondary cosmic rays are formed in the upper atmosphere. Only secondary cosmic rays normally reach Earth's surface.

ments, and subnuclear particles called *secondary cosmic rays* emerge. Scientists can study both primary and secondary cosmic rays by sending rockets and balloons into the upper atmosphere. On the ground, they normally can detect only secondary cosmic rays.

Primary cosmic rays are interesting in themselves because of what they may show about the forces that operate in the universe. But to particle physicists, secondary rays are even more interesting, for they are subnuclear particles that cannot be observed elsewhere in nature. The existence of such particles was proven through cosmic ray studies. Scientists knew the second-

Ways of Observing the Invisible

space. Scientists learned about high-energy radiation of extraterrestrial origin long before space flights were made; some of it is powerful enough to penetrate the atmosphere. When first detected, it was named *cosmic radiation* because it came from the cosmos—the universe —beyond Earth.

No one yet knows where all cosmic radiation originates or how it is produced. At least some of it comes from the sun. The rest comes from much farther away: from supernovas, perhaps, or from the galaxy's core. The forces that accelerate the particles in cosmic radiation and shape their paths are likewise not fully understood.

In space there are various types of cosmic radiation. The type known as "cosmic rays" consists of the nuclei of atoms: that is, protons and neutrons without any orbiting electrons. Most cosmic rays are single protons because a hydrogen nucleus is just one proton and hydrogen is the most abundant element in the universe. The nuclei of other atoms can appear as cosmic rays, however. For instance, a cosmic ray may be a helium nucleus, two protons and two neutrons, which is just like an alpha particle; the only difference is that it is traveling faster than an alpha particle and has a different origin.

Cosmic rays in the form of nuclei are called *primary cosmic rays*. These rarely reach Earth's surface, for when primary cosmic rays enter the atmosphere, they usually collide with the atoms of the atmospheric ele-

Ways of Observing the Invisible

ary rays must come into being as the result of high-energy collisions because the lives of these rays are measured in millionths of a second. There would not be time for them to travel far before decaying if they merely arrived from space.

If secondary cosmic rays have such short lifespans, how can they be studied? This is done by means of *emulsions* or *plastic detectors*. An emulsion is a special kind of photographic film. It is not used to take pictures of particles with a camera; instead, the particles travel through layers of the film itself. Each particle interacts with some of the atoms in the film, so that when the film is developed, the paths of the particles can be seen through a microscope. A plastic detector works the same way, except that the particle paths are made in plastic material rather than in film, and are revealed by chemical etching. Scientists can tell many things about particles by analyzing these paths.

But although cosmic rays give scientists a source of subnuclear particles to observe, this source is not a convenient one. There is no way to predict exactly what particles will arrive in the form of cosmic rays or what direction they will come from, so experiments that depend on cosmic rays cannot be planned precisely. Not even the energy levels of the colliding particles can be planned, for some cosmic rays have much higher energies than others. Furthermore, experiments with high-energy cosmic rays cannot be carried out on Earth's surface. Though secondary rays do reach the surface, it is impossible to tell what produced them without a record of the collision. Records of the collisions that produce secondary cosmic rays can be obtained by sending up emulsions or plastic detectors in balloons and rockets, but this does not give as much information about particle collisions as heavy, immobile equipment can. Besides, there is no way of knowing in advance whether or not significant collisions will occur in the place where the emulsions are sent. Artificial production of subnuclear particles is therefore necessary.

Subnuclear particles can be produced in laboratories

To learn about the subnuclear particles created in collisions, scientists must have a way to control their experiments. In other words, they must bring about collisions rather than merely observe them. In a controlled experiment, the types and energies of the particles that collide can be planned. Scientists can then set up equipment that will record what happens when a specific event takes place.

Machinery is needed to produce high-energy collisions and to make lasting records of the effects. First, there must be a device called a *source* to separate par-

Ways of Observing the Invisible

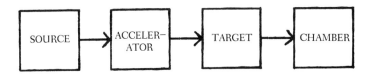

These four kinds of equipment are used in particle physics experiments.

ticles such as electrons and protons from atoms. Then equipment must give these particles energy by causing them to travel faster than they normally do on Earth. An increase in velocity is called acceleration; for example, when a vehicle increases its speed, it is said to be accelerating. Machines that speed up particles are therefore called *accelerators*. In accelerators, electrons or protons—or sometimes whole nuclei—are given high energy by being accelerated to speeds nearly as great as the speed of light.

Next, the actual collisions must be produced. This is done by beaming accelerated particles toward a *target* composed of known atoms. When these particles hit the target, they are moving fast enough to smash the nuclei of some of the atoms and thereby produce new elements or isotopes. In some cases they are moving fast enough to transform nuclear particles within the target's atoms instead of just altering the composition of the nuclei. In these cases, the nuclear particles turn into short-lived subnuclear particles that would not otherwise be observable.

Such particles are, of course, too small to see; and they decay so quickly that they wouldn't be seen even if they were visible. To study them, it is necessary to record their effects. This is done with a device called a *chamber*. Particles created by high-energy collisions in the target have sufficient speed to travel into the chamber before they decay, and photographs of what they do there can be made. Arranging this is not very difficult compared to the difficulty of accelerating particles with which to bombard the target.

Electromagnetic force makes accelerators work

There are several different types of accelerators. Scientists are constantly working to design better ones so that they can accelerate particles to higher and higher speeds. The higher the speed of a particle, the more energy it has. The more energy it has, the more kinds of subnuclear particles can be found when it bombards its target. This is why scientists are still discovering new particles: they are developing accelerators that can give protons and electrons larger amounts of energy.

The energy given to particles in an accelerator is measured in units called *electron volts*. An electron volt is a measurement of an individual particle's energy; it is

Ways of Observing the Invisible

not the same as a volt available from a wall plug or battery. The first accelerators produced particles with energies of thousands of electron volts, but that was not nearly high enough; soon energies were measured in millions of electron volts. A million electron volts was still not very high energy. Today, modern accelerators achieve energies measured in billions of electron volts. The newest ones can reach about 500 billion electron volts, or 500 GeV. Though this may sound like a lot, cosmic rays sometimes have energies of 1,000,000,000 GeV! No machine has yet been built that can reveal all the particles that may exist.

An accelerator is a huge, complicated machine. The two main kinds in use today are *circular accelerators* and *linear accelerators,* but not all the accelerators of each kind work in the same way. Understanding the ways accelerators work requires knowledge of more scientific principles than can be explained here. Basically, however, acceleration is accomplished with electromagnetic force. Only the charged components of atoms—that is, protons or electrons—can be accelerated, although sometimes neutrons bound to protons by nuclear force can be accelerated along with protons. These charged particles are pulled forward through the use of electric current, which produces magnets in circular accelerators and charges thin plates of metal in linear accelerators. Because of the design of the machine, the particles speed

Particles in accelerators travel inside tubes like this. The picture shows only a small section of the tunnel, which continues on for a long distance.

up when they reach gaps between the magnets or charged plates in their path. Once they have speeded up, their momentum keeps them going, so that they speed up still more every time such a gap is reached. They must pass many gaps before they have been accelerated enough to bombard a target. Their paths must therefore be extremely long, many times longer than a football field.

Electromagnetic force controls the direction in which the particles travel as well as their speed. In some circular accelerators they spiral out from the center in widening circular paths; in others they keep traveling in the same circle. The shape of the path is determined by electromagnetism, which can make charged particles move in curves. Circular accelerators are more economical than linear accelerators because, since the particles move round and round in circles, longer paths can be obtained in the available space. In linear accelerators the paths are straight, so the machines themselves must be as long as the distance the particles travel.

Linear accelerators have advantages, however. For instance, it is easier to put electrons or protons into a linear accelerator than to put them into a circular accelerator. It is also easier to get high-energy particles out of linear accelerators and to control the way they bombard their target. In addition, linear accelerators are more efficient than circular accelerators in certain cases, because in

This is the circular accelerator at Fermi National Accelerator Laboratory near Chicago. The circle, which is called a ring, is almost two kilometer across, and is the largest in the United States.

This is the linear accelerator at Stanford Linear Accelerator Center near San Francisco. It is over three kilometers long.

The Subnuclear Zoo

circular accelerators some particles are "lost" in the form of radiation before reaching the target.

Although most accelerators produce beams of high-energy particles that hit a target of normal, unaccelerated atoms, it is also possible to have machines in which two beams of accelerated electrons or protons collide with each other. Collisions of far higher energy can be produced by this method than through the use of a single beam. However, such accelerators are not suitable for all types of experiments.

Effects of particle collisions are recorded in chambers

When accelerated particles collide with the atoms in a target or with each other, subnuclear particles are produced. These particles cannot be seen, or even photographed, directly. It is therefore necessary to make photographs of their effects. This can be done because when they pass through machines called chambers, they leave tracks, or else decay into other particles that leave tracks. By analyzing pictures of these tracks, like the one on page 58, scientists can tell a great deal about the behavior of subnuclear particles.

The most common type of chamber used in high energy physics experiments is called a *bubble chamber*. This machine has a large container filled with liquid.

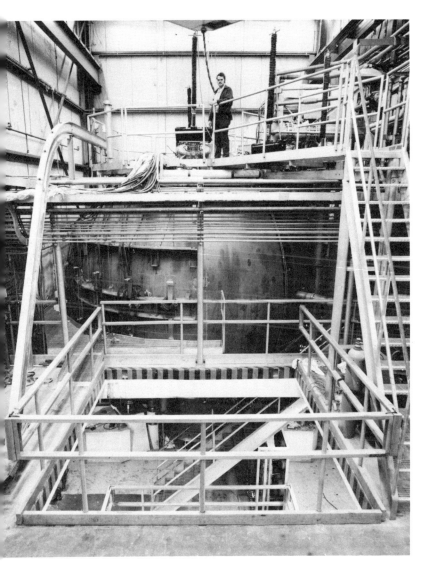

Bubble chambers like this one make particle tracks visible. The large tank is filled with liquid. Particles traveling through it leave trails of tiny bubbles that can be photographed.

Though hot enough to boil, the liquid does not boil because it is under pressure. The pressure is released just as the particles enter the chamber. As the liquid starts to boil, tiny bubbles form along paths the particles take going through the chamber; this leaves a track that can be photographed. After the photograph has been made, the liquid is pressurized again in preparation for more particles.

Various kinds of liquids are used in bubble chambers, depending on the kind of atoms physicists are using for their experiments. Sometimes the atoms in the liquid itself are the targets accelerated particles hit. At other times, the accelerated particles hit a target of solid material such as a thin piece of metal, and subnuclear particles formed by that collision are what enter the bubble chamber.

Another type of chamber, which is more suitable for certain experiments, is called a *spark chamber*. The tracks photographed in spark chambers are trails of electric sparks instead of trails of bubbles. Spark chambers are filled with gas rather than liquid and most types contain a series of charged metal plates. As a particle moves through the chamber, it causes sparks to jump from plate to plate. One advantage of spark chambers is that they do not need to be turned on unless it is known that a desired particle is coming through. Radiation counters can be set to turn on the electricity auto-

The machine in this picture is a spark chamber. Particles make sparks as they pass through it, and the pattern of sparks can be photographed to record the paths those particles followed.

matically when a certain particle approaches. The camera can be triggered automatically, too, so that film is used only when something worth photographing happens. Some types of spark chambers have wires instead of metal plates, and do not use film at all; they record the positions of the sparks in the memory of a computer rather than in a picture. This makes it easier for scientists to study the data.

Scientists learn about particles from track photographs

Scientists study the tracks made in emulsions and chambers when accelerated particles collide with target atoms. But how do they know which particles made the tracks? Tracks, after all, look very much alike. To a person who is not a physicist, one photograph of particle tracks seems just like another except for having a different pattern of lines. Why should anyone think that the lines are made by different kinds of particles? Why should physicists get tremendously excited—as they do—over certain photographs that are apparently similar to all the rest?

The pattern of lines is the key. Physicists are able to predict the directions in which familiar types of particles will travel after collisions, how far those particles can go before decaying, and what they will turn into

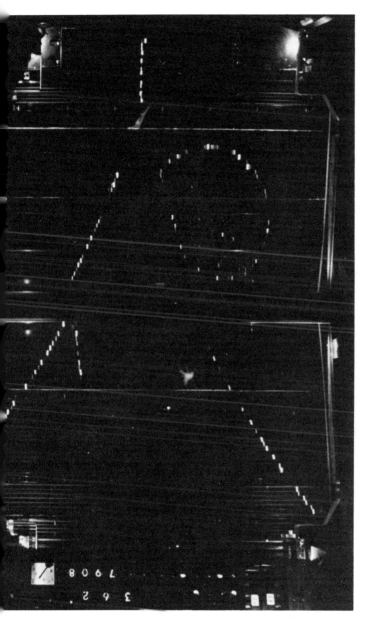

This picture shows the trails of sparks left by particles passing through a spark chamber.

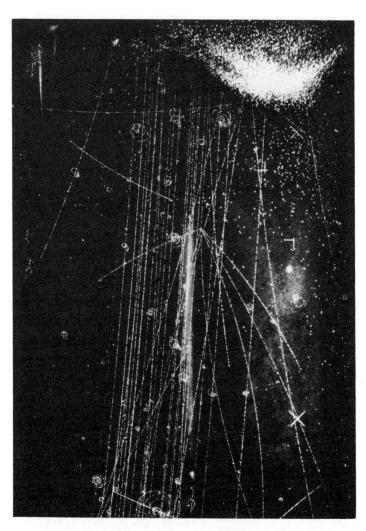

This is a typical bubble chamber photograph. The straight lines running from top to bottom are tracks left by a beam of particles that passed through the chamber. Near the top of the picture two particles collided, producing subnuclear particles that went off in different directions.

when they decay. Because of this, they are able to identify the individual particles in a track photograph by measuring the lines.

Usually, the photographs obtained in an experiment are meant to prove predictions. Scientists have theories about how certain particles will behave, but they cannot be sure of the theories without analyzing many thousands of track pictures that show the particles actually behaving as they were expected to. Such photographs are evidence that the theories are correct.

Sometimes, however, the pictures do not confirm the theories; they show patterns that do not fit all the predictions about what particles will do. Then scientists see that there are facts not yet known, and they plan more experiments to find out what is going on. This is the most exciting part of particle physics: the discovery of new facts that no one had suspected. Occasionally particles identified by the beginnings of their tracks do not do what they were expected to do in the rest of the photograph, and new theories must be developed to account for their actions. More often, especially when collisions involving higher and higher energies are studied, tracks are seen showing things that no familiar particle could possibly do. Scientists then realize that they are looking at the actions of a "new" particle.

Of course, "new" particles are not really new. They are simply newly discovered. Once scientists believed that comparatively few kinds of particles exist and they did not expect to discover any more. Now they believe there are many subnuclear particles not yet seen. Some scientists have even begun to speculate that there may be an infinite, or unlimited, number of particle types. Whether this is true or not, it will be a long time before physicists have observed all the types of subnuclear particles that experiments can reveal. There is still plenty of interesting work to be accomplished.

The Members of the Zoo

How do particles differ from each other?

It is easy to see how an electron differs from a proton, how a proton differs from a neutron, and so forth. Imagining more than a hundred kinds of subnuclear particles with different characteristics is harder. It might seem that there are very few ways in which invisible blobs of matter and energy can differ. That is not the case, though. Particle types differ from each other not only in properties like mass, but in more mysterious characteristics that determine their actions. This great variety of characteristics is what makes scientists picture subnuclear particles as members of a "zoo."

Subnuclear particles are not classified the way animals in a zoo are, however. The way animals act does not have anything to do with their classification into species; scientists can separate life forms by type with-

THE SUBNUCLEAR ZOO

out taking their behavior into account. Every living creature differs from every other as far as its actions are concerned. Particles of the same type, on the other hand, do not differ in their behavior. And it is the way they behave that tells what type they are. If scientists observe subnuclear particles that act in different ways, they know these particles are of different types. That is why more and more new types are constantly being discovered.

To be sure, within certain limits there is variation in the behavior of identical particles. For example, though nuclei in the atoms of each unstable element decay, there is no known way to predict when any particular atom's nucleus will decay. Predictions about radioactive decay are expressed in terms of how percentages of particles will behave, not how specific ones will behave. This is the reason scientists speak of an unstable element's half-life instead of the whole lifetime of any one

Not all identical particles decay at the same time. They do not all decay into the same subnuclear particles, either.

atom. Also, the kind of decay that will occur—alpha, beta or gamma—must sometimes be predicted in terms of percentages. In some cases, not all atoms of an element decay in the same way, even though they are all composed of particles that are apparently alike. The same principle applies to decay of the short-lived subnuclear particles. Scientists know how percentages of each particle type will act, rather than how the individual particles themselves will act.

Knowledge of this sort is called *statistical* knowledge. Just as insurance companies can predict how many people out of a thousand will die at a given age, but not which people, physicists can predict how many particles will decay in a given time without knowing which particles. Such facts are statistical facts. The statistics are more exact for particles than for people, because unlike people, particles cannot make any choices that affect what happens to them.

Since knowledge of particle behavior is statistical, it cannot be said that all particles of the same type always act in exactly the same way. Nevertheless, types of subnuclear particles are identified by measurements of their actions. Scientists can determine the characteristics of the types only by observing what those types do.

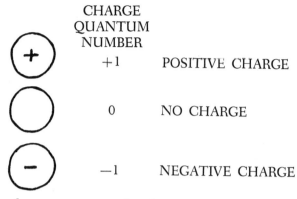

The charge quantum number of a particle measures the amount of electricity in that particle. The sign before the number tells whether the electricity is positive or negative.

Different particle types are defined by quantum numbers

The different types of particles are defined by what are called *quantum numbers*. A quantum number is a measurement of some characteristic of the particle—charge, for example. All particles that have the same set of quantum numbers and the same mass, or amount of matter, are of the same type.

There are special units of measurement for each particle characteristic, and a quantum number tells how many units of that characteristic a particle has. For instance, the unit of charge is 1. A particle with a positive charge has a charge quantum number of $+1$, one with a negative charge has -1, and one with no charge has 0. It is not possible for a particle to be partly charged or to possess any other quantized characteristic in amounts

The Members of the Zoo

other than multiples of established units. Determining quantum numbers is therefore more like counting than like measuring.

Most of the characteristics represented by quantum numbers are not easy to describe in words or to picture in physical ways. Scientists usually make statements about them in mathematical terms, using symbols. When they do use words, the words chosen are merely "name tags" that make it possible for physicists to communicate with each other. Such names are at best analogies and do not tell what particles are actually like, just as picturing a miniature solar system does not tell what a nucleus surrounded by electrons really looks like.

One quantum number, for example, represents *spin*. Many particles act as if they were spinning on an axis in

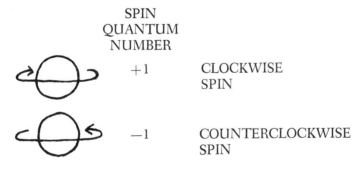

Photons have a spin quantum number of +1 if they are spinning in a clockwise direction. They have a spin quantum number of −1 if they are spinning in a counterclockwise direction.

a way similar to the rotation of the earth. But of course, no one has seen a particle spinning, and no one can be sure that particles actually do spin. All that scientists know is that a property something like spin must be taken into account to make the mathematical predictions of particle behavior work out right. This property can be measured, and since the equations do work, its importance has been proven. "Spin" is as good a name for it as any.

Some of the other names are more misleading to people who are not physicists, although the physicists themselves understand what is meant when these words are used. One such name is *strangeness*. Around the middle of this century, scientists discovered some particles that behaved in strange ways compared to the particles they already knew about. They began referring to them as "strange" particles. Before long, this name had become firmly attached to the characteristic that had seemed strange at first—so firmly that "strangeness" became its official name. When a physicist now speaks of a strange particle, he means one that has a strangeness quantum number other than zero. And indeed, there is no concrete, graphic word that could be used for the property this quantum number expresses.

Yet there are qualities even more peculiar. Knowing that these qualities are indescribable, scientists have had fun naming them. Among the newest names for quan-

The Members of the Zoo

tum numbers are *charm, color* and *flavor.* Not all of these properties are yet known to exist; charm has been observed, for instance, but color is still completely theoretical. Scientists have begun to talk about "red," "white" and "blue" particles because they believe certain particles not yet discovered have three possible values for the quantum number "color." This does not mean anyone supposes such particles are really colored red, white and blue, or any color at all. It simply means that imagining them as if they had colors is useful for remembering that they differ from each other.

Not all quantum numbers are applicable to all types of particles. The property called color, for example—if it is proven to exist—will not apply to electrons. Physicists use specific sets of quantum numbers for specific classes of particle types. The scheme of defining particles is extremely complicated and confusing, not only to nonscientists, but to scientists themselves. Physicists are working to acquire knowledge that will make this scheme less complicated.

All particles have antiparticles

One of the things physicists have learned about particle types is that all particles have *antiparticles*. These antiparticles are types with the reversible characteristics of the corresponding particle reversed: for instance, the positive quantum numbers changed to negative ones. Charge is one example of a reversible characteristic. Therefore, for each type of particle with a positive charge there exists an antiparticle with a negative charge, and vice versa. Such particles and their antipar-

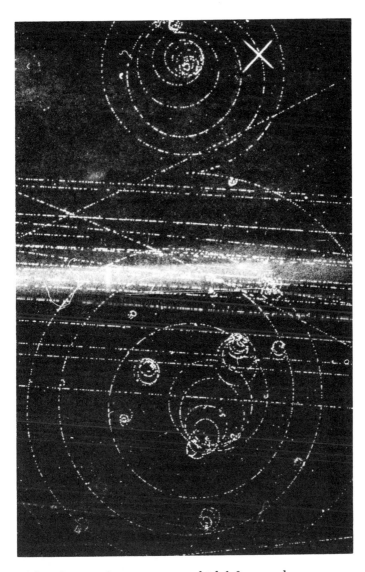

In this photograph a gamma ray, which left no track, materialized into an electron and a positron. The bottom spiral was made by matter; the top spiral was made by antimatter.

ticles are identical except for charge and whatever other reversible charactcristics they possess. A particle with no reversible characteristics, like the photon, is considered to be its own antiparticle.

Protons and electrons are charged particles; therefore protons have antiprotons and electrons have antielectrons. Protons and electrons are not antiparticles for each other because their non-reversible characteristics are not identical. The mass of the proton is far greater than the mass of the electron, and there is nothing reversible about mass—it is a characteristic with no "opposite." The mass of a particle and the mass of its antiparticle are always exactly the same.

The first antiparticle discovered was the antielectron, which is just like an electron except for having a positive charge and some other reversed quantum numbers. It was named the *positron*. But scientists decided that to have a separate name for every type of antiparticle would be too confusing. Antiprotons are therefore named simply "antiprotons," and the same system is followed for all the rest in cases where names are used instead of mere symbols.

Antiparticles are not components of ordinary matter. They are not found in atoms, at least not in our part of the universe. Atoms made of antiparticles would be not matter, but *antimatter,* and no one knows whether or not objects made of antimatter exist anywhere. Nevertheless, antiparticles do exist outside atoms. They come into being during some forms of beta decay, and also during decay processes that result from high-energy particle collisions.

Why are so few antiparticles found in nature? In one sense that is an easy question to answer. When particles meet their antiparticles, they annihilate each other—in other words, all their mass is transformed to energy by the interaction. A positron emitted during beta decay cannot exist very long before it encounters an electron and the two are converted into pure energy. Antiparticles created in laboratory experiments do not last long either, for the same reason. And antiparticles in cosmic radiation do not get far into the atmosphere without being annihilated by their counterparts.

It is also easy to say why we do not live in a world made up of half ordinary matter and half antimatter. Such a world could exist only a few instants before turning into pure energy—we would not be here to observe it. But in another sense, the question of why our world has more matter than antimatter is something scientists do not understand. Why should part of the universe start out with more of one kind of matter than the other? That is a mystery that cannot yet be solved. Perhaps the universe as a whole has equal amounts of each kind; some scientists believe that there may be galaxies composed of antimatter. If there are, we could not tell by observing them through telescopes, because the light given off by antimatter would be just the same as the light emitted by ordinary matter.

Stories have been written about what would happen if things made of ordinary matter met things made of antimatter, and they are fiction. The concept of antimatter itself is not fiction, however, nor is it mere theory. Antiparticles are as real as any other short-lived particles that have been observed in high energy physics experiments.

Particle types are divided into families

The various types of particles and their antiparticles are divided into families according to their roles in the structure of matter. It is not possible to make a simple list of these families because some of them overlap. Scientists use different classifications for different purposes. In addition, as they learn more about subnuclear particles, they find new and better ways of classification. For these reasons not all charts of particle families are alike, and no existing chart can be considered final.

It is easier for scientists to name particle types and families of types than to name the mysterious characteristics represented by the quantum numbers. Some families are named after the scientists who discovered things distinguishing those families. Others are given names derived from Greek words or letters. This is also

The Members of the Zoo

Families		Particle	Antiparticle
Boson	—	PHOTON	Same particle
Fermion	LEPTON	NEUTRINO ELECTRON MUON	ANTINEUTRINO POSITRON ANTIMUON
Boson	MESON	PION KAON ETA MESON	ANTIPION ANTIKAON Same particle
Fermion	HADRON / BARYON	PROTON NEUTRON LAMBDA HYPERON SIGMA HYPERON CASCADE HYPERON OMEGA HYPERON	ANTIPROTON ANTINEUTRON ANTI-LAMBDA HYPERON ANTI-SIGMA HYPERON ANTI-CASCADE HYPERON ANTI-OMEGA HYPERON

This chart shows some of the best known particles that have been found and the families they belong to. There are over a hundred particles not included in the chart, many of which are identified by mathematical symbols instead of names. Physicists are constantly finding new particles, and often they cannot tell what families they belong to until many experiments have been done.

true of the particle types, but many particles are given just letters or letters combined with their family names. There are now too many types to give them all unique letters, so some members of the zoo are identified by letters plus their quantum numbers.

The Subnuclear Zoo

There is no need for anyone but physicists to know all the many types of subnuclear particles and their characteristics. The significance of the quantum number differences, in many cases, can be explained only by complex higher mathematics in which equations must be used instead of words. However, it is interesting to know about some of the most important families of particle types.

One way to classify particles is to divide them into two groups: *leptons* and *hadrons*. Leptons do not participate in strong interactions, which is another way of saying that they are not affected by the strong nuclear force. Electrons and neutrinos are in the lepton family. Hadrons do participate in strong interactions. Protons and neutrons, naturally, are in the hadron family, since it is strong nuclear force that holds them together. There are other particles in each of these two families. The only observed particle that is not in either family is the photon.

The hadron family is divided into subclasses called *mesons* and *baryons*. Mesons are particles that carry strong nuclear force, while baryons are particles that are affected by it. This means protons and neutrons are baryons. There are further divisions within the baryon subclass; protons and neutrons belong to one of them.

Physicists have another way of dividing particles into families: all types are either *fermions* or *bosons*. This

The Members of the Zoo

distinction depends on spin quantum number and is useful because it tells what kind of mathematical rules the members of these two families follow. However, no type of observed particle (except the photon) belongs only to the fermion family or only to the boson family; it has its place in either the lepton family or the hadron family also.

Fermions are particles affected by forces, so baryons, for example, are fermions. Bosons are particles that carry forces, which means that mesons are bosons. It also means that photons are bosons, since photons carry electromagnetic force. Scientists believe that there are bosons that carry gravitational force and weak nuclear force, and have named them "gravitons" and "W particles." So far, though, these two bosons have not been observed.

Particle types are not put into families just for convenience; there is a more important reason. Knowing the families a particle belongs to is an essential part of discovering how that particle acts. In the first place, remembering all the details about each individual type of particle would be hard to do. It is much easier for scientists to learn facts that apply to whole families of types and then to remember each particle's families. Furthermore, once enough has been learned about a "new" particle for its families to be known, then unobserved facts about it can be predicted. This is so because particles must "obey the rules" of their families. Knowledge of these rules makes it possible to design experiments for verifying predictions about particle behavior. If there were no rules, scientists would have no basis for such predictions, which would mean they could not identify particle tracks by comparing actual results of collisions with expected results.

There are rules for particle interactions

The rules that are followed when particles interact are among the most important things scientists discover. Knowing the rules helps scientists learn about particles, and furthermore, observing particles is a method of learning about the rules. Studying particles would have little meaning if there were no pattern to their actions. The rules are important not so much because they are used to predict particle behavior—although this is true—but because these rules are statements about how the universe works. Finding out how the universe works is the real goal of science.

Science is the discovery of patterns and principles that are called natural laws. These laws mean much more than "laws" of the sort made by human beings. They are built-in ways in which the universe operates, and without them, nothing in the universe would hold together. Knowing what to expect of the universe is the truly significant thing; knowing what to expect of particles is simply a means to that end.

Although the laws of the universe cannot be changed by scientists, beliefs about the laws can and do change. Sometimes what is believed to be a law isn't as firm a law as was originally thought. When scientists discover laws, they cannot be sure that the laws will apply in all circumstances. If they find circumstances in which known laws do not hold, they do not throw away the laws entirely. A law does not cease to be a law just because there are cases to which it does not apply. Such cases merely show that the pattern of natural laws is more complicated than people had believed it was, and that the complications have to be understood before simple explanations can be seen.

Once, it was believed that the amount of matter in the universe could never change and that the amount of energy could never change either. When scientists learned that matter and energy are interchangeable forms of the same thing, it looked as if the old laws about matter and energy did not hold. But the laws were very important to science. They explained a great deal about things that had been observed in the past. So the laws were not thrown away; they were reinterpreted. This was done through mathematics that showed that the total amount of mass-energy in the universe does not change. The mathematics had to take into account complicated facts about transformation of matter to energy and vice versa, yet the result was a simplification. One law replaced two separate ones.

The law that mass-energy must be conserved is an example of a *conservation law*. It tells scientists that in any interaction between particles, the total amount of mass-energy involved cannot change. If a collision between particles causes some energy to be lost, then particles with greater mass must come into being. If massive particles decay into less massive ones, then more energy must appear.

There are quite a few other conservation laws that apply to interactions between particles. Charge must always be conserved, for example. A positive particle and a negative particle can combine into a neutral one; "neutral" simply means that the amounts of posi-

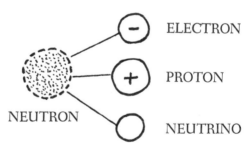

Charge is conserved after beta decay. The amounts of positive electricity and negative electricity in a neutron are equal. The total amounts in the particles emitted during decay are also equal.

The Members of the Zoo

tive and negative electricity are equal. But a neutral particle cannot decay into particles with a greater total positive charge than negative charge, or vice versa. None of the positive or negative electricity can be transformed into the opposite kind. Whenever particles decay or come into being, the amounts of positive and negative charge involved remain the same.

Other things that are conserved in most particle interactions are momentum, spin, strangeness, and so forth. Every quantum number represents a property that is conserved. When particles interact, the sum of their quantum numbers for each given property does not change.

Not all the conservation laws apply to all types of interactions, however. For instance, strangeness is not conserved in weak interactions. Neither is a property called *parity*, which has to do with the mathematical description of the directions in which interacting particles move. Scientists were surprised to discover that there are cases to which the law of parity conservation does not apply. It had been thought that in nature there is no distinction between "left" and "right" or between "up" and "down," and that in any type of interaction particles are as likely to travel in one direction as in another. But in beta decay, a weak interaction, most of the particles travel in a predictable direction, which shows that a "mirror" world could be distinguished from our world. This exception to the law complicated things for physicists. They could not throw the law away because it explained too much about other interactions. So again, they tried combining several separate laws to see if one general one could cover all cases. They found that it could if special cases involving antimatter were included in the calculations.

It is unlikely that scientists of today know about all the conservation laws and other rules that affect the behavior of particles. New ones are still being discovered, just as new particles are. But the known laws are extremely important. They are what enable scientists to understand track photographs; the patterns of lines in the photographs make sense only to people who have learned rules for interpreting those patterns. Since members of the subnuclear zoo are invisible in themselves, they cannot be studied except through their actions—and identifying these actions depends on knowledge of the rules.

Some Mysteries Still to be Solved

Physicists are searching for simpler patterns

Scientists have learned so much about the types and behavior of particles that their work has become extremely complicated. Sometimes it looks as if they are trying to make particle physics more complicated every year. There are so many details that it's almost impossible for anyone to keep track of them all. It is easier to start wondering whether further study is worth the effort. What is there to gain by finding more particles, anyway?

Physicists themselves ask this question frequently. If the only goal of their work were to collect more and more details about an ever-growing collection of confusing particles, it might not be worth either the effort or the expense of building new accelerators.

But gathering details is not the only goal. What scientists want to do is make sense out of the confusion. They know that the old, simple patterns once taught in physics are not complete and that newly discovered facts about particles and interactions cannot be ignored. All the same, they are searching for new patterns that will make these facts simpler to explain.

There are two main things physicists are hunting for. First, they are trying to identify *fundamental particles* of which all other particles are made. Discovering so many members of the subnuclear zoo has been exciting, in one way. But in another way, it has been upsetting to find such a great variety of different particles. There are now more known types of subnuclear particles than there are elements. When scientists first learned that the atoms of the different elements are made of electrons, protons and neutrons, they were glad. It had not seemed reasonable for there to be a large number of different building blocks for matter; it was much more logical for there to be just a few fundamental particles that fit together in different combinations. Electrons, protons and neutrons were assumed to be fundamental until knowledge about subnuclear particles complicated the picture.

For a long time, scientists believed the subnuclear particles they detected were fundamental, or "elementary," too. Many books describe these particles in such terms. Recently, however, high energy physicists have begun to believe that most of the observed particles are

Some Mysteries Still to Be Solved

not really fundamental. They have been developing theories about how these particles may be made of still smaller ones that have not yet been discovered. If this is true, the pattern revealed by particle physics will be resimplified. Physicists are therefore working hard to learn whether or not any truly fundamental particles exist.

The second thing physicists are searching for is a simpler way to explain the forces that cause particles to interact. Several basic forces are known to exist in the universe: gravitational force, electromagnetic force, strong nuclear force and weak nuclear force. Some scientists think other basic forces may also exist, such as a superweak force or a semistrong force. It looks as though the study of forces is getting more complicated, just as the study of particles is. However, most scientists believe it's unlikely that the universe has some specific number of forces that are unrelated to each other. They would like to explain all interactions in terms of a single force that manifests itself in various ways. The effort to do this is known as the search for a *unified field theory*.

These two goals—the discovery of fundamental particles and the development of a unified field theory—are the most important aims of high energy physics. Current experimentation is important because it may someday lead to the achievement of these goals.

Are quarks the fundamental particles?

Most physicists today believe that the members of the hadron family—baryons and mesons—are probably made of smaller particles called *quarks*. There appear to be too many types of hadrons for them all to be fundamental; it seems as if it should be possible to account for their differences by explaining how they are structured. The quark theory provides such an explanation. However, it is still only a theory. It has not been proven because no one has observed any quarks. Scientists are not sure they will ever be able to observe any.

Are protons, neutrons and other hadrons fundamental particles?

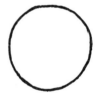

Or are they composed of smaller particles called quarks?

Scientists are not yet sure. Some of them think that even if quarks are found, there could be still smaller subnuclear particles inside them.

Some Mysteries Still to Be Solved

At first, physicists who speculated about quarks suggested three types, plus three types of antiquarks. The quantum number defining the types is called *flavor*, though of course it has nothing to do with how things taste. The three flavors were, in fact, named "up," "down" and "sideways" instead of being given names that sound like real flavors. These three were thought to be enough to explain all the differences between various types of hadrons. Now, however, a fourth flavor called "charm" is needed to explain the characteristics of certain hadrons. Some physicists believe there must be two more flavors in addition to charm, which would make a total of six.

Besides the different flavors, quarks are now believed to have different colors. The reason is that quarks must be fermions, and according to the rules all fermions obey, two with the same set of quantum numbers cannot be in the same particle. Yet many hadrons—protons and neutrons, for example—do have to contain two quarks with the same flavor. So if quarks exist, there must be differences between them besides flavor, and the quantum number for color is used to describe these differences. It must be remembered that "color" is only a name tag; no one thinks quarks are actually colored. Furthermore, the property defined as color does not produce any observable distinctions between hadrons. This means each hadron has to contain the same mixture of "colored" quarks. Quark flavor, not color, makes hadrons unlike each other.

The quark theory is more than a purely imaginary idea because it enables scientists to make predictions about particle behavior that match experimental results seen in track photographs. However, no quark has yet been produced in a high energy collision. Some scientists believe that far more energy than present accelerators can achieve would be required to separate the quarks in a hadron and that if such an amount of energy were used, more quarks would materialize, combining into other particles before there was time to record evidence of the quarks' separate existence. If this is so, more powerful accelerators will never make it possible to observe quarks. But further evidence for the theory can still be found by proving that hadrons act as if they were made of quarks. Some such evidence has been obtained from experiments in which neutrinos, or sometimes electrons, collide with hadrons such as protons or neutrons.

If quarks are the building blocks of hadrons, the number of fundamental particles is much less than if hadrons themselves are fundamental. Nevertheless, there are still a great many fundamental particles even if quarks exist. In addition to quarks, there are leptons, which unlike hadrons are not thought to be composed of smaller particles. Four types of leptons are known to

Some Mysteries Still to Be Solved

exist: electrons, muons and two kinds of neutrinos. There is evidence that a fifth type may exist, and because leptons seem to come in pairs, physicists would not be surprised to find a sixth one also. Six types of leptons and six flavors of quarks would be a nice, neat arrangement making a total of 12 fundamental particles. But if each flavor of quark comes in three colors, the total is 24; and since every particle has an antiparticle, the 24 has to be doubled. That makes 48. The idea that there may be as many as 48 fundamental particles is discouraging to scientists; it raises the question of whether there is any simple pattern in the structure of matter at all.

Physicists are divided in their opinions about this question. Some think leptons and quarks are indeed the basic building blocks, even though there may be a lot of types. Others think leptons and quarks themselves may be made of still smaller subnuclear particles. A few believe there may be no end to the search; they say that perhaps particles are like a series of seeds, one inside the other, with no such thing as a "smallest" seed that has no smaller one inside. It will be a long time before anyone can prove which of these ideas is true. Perhaps no one ever can, but scientists will keep on trying.

The Subnuclear Zoo

Is there only one force in the universe?

Another goal scientists will continue to work toward is the unification of knowledge about forces. Physicists are not yet sure whether all universal forces can be explained as different aspects of one force, but most agree that the search for a unified field theory is worth pursuing. This search is not an easy one, and it is not expected to be finished soon. Researchers do not yet know enough facts to tie together all that has been observed about gravitational force, electromagnetic force, strong nuclear force and weak nuclear force. So far, they have been unable even to guess many details about the underlying unity they suspect exists; that is why they are still searching for the theory itself instead of using it to make predictions that can be verified in experiments. After a theory has been developed, the next step will be to prove it.

A theory that unified the forces would be a "field theory" because scientists believe that all forces are carried by *fields*. A field can be thought of as a stream or current of the bosons that carry a force. For example, electromagnetic force is transmitted from one charged particle to another by a stream of photons called an electromagnetic field, or a photon field. A similar process exists involving strong nuclear force and mesons and

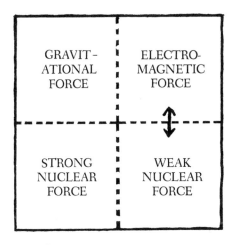

Are the forces in the universe really separate forces or should the dotted lines be removed from this chart? Scientists are trying to find out. So far, they have discovered a link between electromagnetic force and weak nuclear force.

is believed to exist with gravitational force and gravitons as well as with weak nuclear force and W particles.

The first step in developing a unified field theory is to discover facts about how the known forces are related. This is the stage where physicists are now. At present, most of the information they have gathered is about how electromagnetic force and weak nuclear force are alike. A striking similarity between the two is that both act on leptons and hadrons, unlike strong nuclear force, which acts only on hadrons, and gravitational force, which acts on all particles including photons. There are some mathematical similarities, also.

However, the apparent differences between electromagnetic force and weak nuclear force are even more striking than the likenesses. For instance, electromagnetic interactions within a nucleus take place ten billion times faster than any weak interactions there. Furthermore, electromagnetic force has an infinite range, while weak nuclear force has a range much smaller than the diameter of a proton. Another difference is that the law of conservation of parity can be violated in weak interactions, but never in electromagnetic interactions. And finally, electromagnetic interactions proceed through a neutral current: in other words, the bosons that carry electromagnetic force are uncharged. This was thought to be untrue of weak interactions until recently.

The first clue physicists had in the attempt to eliminate some of these differences between the two forces involved the neutral current. They knew the current that carries electromagnetic force is neutral because when charged particles interact through electromagnetism, the amounts of positive electricity and negative electricity in them do not change. This means the bosons transmitting the force—which are photons—do not carry any positive or negative charge; they are uncharged, or neutral. In weak interactions, on the other hand, charge is usually transferred from one particle to another. When an electron and a neutrino interact through weak nuclear force, the electron becomes a neutrino and the

Some Mysteries Still to Be Solved

neutrino becomes an electron, something that could not happen unless charge was carried between the two by the current that produces the interaction. So if the current is composed of W particles, those particles have to be charged. Some must have positive charges and others must have negative charges to account for the observed results of weak interactions. But mathematical calculations made physicists think there should be some uncharged W particles, too. Uncharged W particles would make a neutral weak current like the neutral electromagnetic current, so the physicists were eager to learn whether or not such a current exists.

They could not simply hunt for uncharged W particles. No W particles of any kind have been detected yet; today's accelerators do not have enough energy to create the conditions that would be needed to produce observable ones. To find a neutral weak current, scientists had to plan experiments where the results of weak interactions could be observed. They knew many of these interactions would produce leptons to which charge had been transferred by positive or negative W particles. However, they hoped that by searching long enough they would find some cases where no charged leptons were created.

Weak interactions are not very frequent, compared to other interactions—that is why the weak nuclear force is called "weak." The neutral weak current ex-

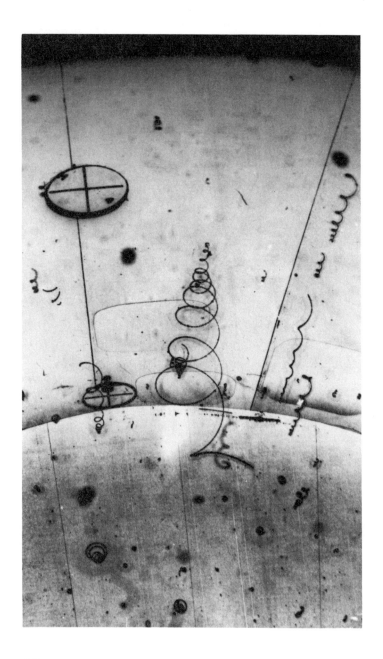

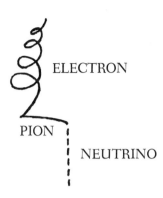

Scientists have looked at hundreds of thousands of photographs like this one to find evidence for neutral weak currents. Here, a neutrino came up from the bottom of the picture, leaving no track. It collided with a proton and the collision produced a particle called a pion, which went from right to left and made the short curved track in the center of the photo. The pion decayed, and the final product was an electron that left a corkscrew track. The fact that no muon came out of the collision shows physicists one similarity between electromagnetic force and weak nuclear force.

periments were hard to perform, and at first they were not successful. Then new accelerators that achieved higher energies became available, and physicists tried again. They did experiments with neutrinos, which do not become involved in any significant processes except weak interactions. When a neutrino interacts by means of a charged weak current, a negatively charged lepton called a *muon* is produced, but a neutral weak current cannot produce any muons. Although checking for the absence of muons may sound easy, it was not easy at all. One group of experimenters had to look at almost 300,000 track photographs to see just 742 cases of neutrino and antineutrino interactions. But of those 742 cases, 166 did not result in muon or antimuon production. This was strong evidence that neutral weak currents do exist, and other groups of experimenters got similar results.

In the same experiments, weak interactions involving neutral currents did not seem to violate the law of parity conservation as often as charged weak interactions do. Since electromagnetic force never violates that conservation law, this suggests that under some conditions weak nuclear force is not as different from electromagnetic force as was once thought. Moreover, scientists have begun to reconsider the biggest difference of all between the two forces, the difference in strength. Perhaps weak nuclear force isn't always as weak as it has seemed to be.

Weak nuclear force is considered "weak" because the probability that it will cause interactions between particles is not large. But recently, there has been some evidence that the probability of weak interactions occuring may increase when the energy of the particles increases. Some physicists think there may be a point, at a very high energy, where the probability of a weak interaction taking place is greater than the probability of an electromagnetic interaction. If that is true, then the weak force and the electromagnetic force do not really differ in strength; they simply operate differently, depending on the energy of the particles upon which they act. However, an accelerator about a thousand times more powerful than today's largest one would be needed to test this idea.

The discovery of neutral weak currents has forged an important link between electromagnetic force and weak nuclear force. Even more importantly, it has caused physicists to become interested in understanding that link. It has made them feel that finding facts to support a unified field theory may not be an impossible task. Already, theories about distant astronomical objects point to possible similarities between gravitational force and electromagnetic force. Scientists have also begun to look for links between strong nuclear force and the other forces. In time, they may learn that there are no true separations between forces, and that just one kind of power causes all physical interactions. This would be a very large step in man's understanding of the universe.

What is there to gain from particle physics?

The goal of high energy physics is to find patterns in the universe: a pattern of fundamental particles, a pattern that unifies all forces, and someday perhaps even a way to merge those two patterns into one. Physicists feel the effort and expense of their work is worthwhile. But why? What difference does it make?

It will make a tremendous difference as far as understanding the universe is concerned. Knowledge of how matter is structured is an essential part of understanding the universe. So is knowledge of what forces keep the physical processes of the universe going. Furthermore, such knowledge may lead to comprehension of many cosmological mysteries. The study of matter and antimatter is related to questions about how the universe began. The study of neutral weak currents is thought by some scientists to be related to the question of whether the universe will keep on expanding forever. Neutral weak currents may also have something to do with the question of why stars explode into supernovas. These are only a few examples of the mysteries high energy physics may help to solve.

Human nature makes people want to solve such mysteries. There have always been men and women who have wondered about the universe; probably there always will be. People are curious. Many cannot be satisfied unless they do everything possible to investigate things they are curious about. That is why they explore new lands and why they become scientists. No reason is necessary; it is simply the way human beings are.

Some Mysteries Still to Be Solved

Is curiosity alone enough to justify high energy physics? Experimenting with accelerators is expensive. People cannot undertake it by themselves whenever they want to. Once, physicists could establish laboratories of their own and make important discoveries without consulting anybody else. That is no longer true; the equipment needed to make new discoveries in particle physics costs billions of dollars to build, as well as a lot of energy to operate. It has been proposed that accelerators larger than those already built should be international projects because they would cost more than any single country could afford.

Although many people feel that understanding the universe is worth this cost, others believe that there ought to be practical benefits. They ask what practical value there is in particle physics. What can be learned that will help Earth's citizens? Can particle physics lead to the discovery of new energy sources? Can it make possible the invention of things often described in science fiction, like antigravity or starships? Can it make life better for anyone?

The only answer to these questions is that nobody knows. Scientists do not say that particle physics will bring about important new inventions. Most of them believe that eventually it probably will, but they cannot tell what kind of inventions. They cannot point to any specific benefits for Earth's people because they have no way of predicting in advance where discoveries in high energy physics will lead. They do not have enough facts yet to make such predictions.

However, there is one thing scientists do know. They know that in the past, whenever anyone has made an effort to understand more about the universe, practical benefits have been the result. The men who began studying electricity did so simply because they were interested in it. They did not imagine any of the things that can be done with electric power, and they certainly did not predict that it would improve the lives of millions of people. Later, those who discovered radioactivity did not see any practical value in that, either. They did not know it could be used to save the lives of cancer victims, as often happens today. The scientists who first studied the structure of atoms did not foresee the constructive uses of nuclear power. Today's particle physicists cannot foresee the things some future generation may be able to do with the discoveries they now are seeking.

Physicists know only that their work is an investment in the future. All past investments of that kind have paid off. High energy physics is unlikely to be an exception, and since its goals concern the whole universe, it may well prove to be the key that unlocks the universe for mankind.

Index

Accelerators, 45–52, 81, 86, 91, 93, 95, 97
Alpha decay, 32–33
Alpha rays (alpha particles), 32–33, 37, 41
Antimatter, 69–72, 80
Antiparticles, 68–73, 87, 94
Atomic number, 10, 33–34
Atoms, structure of, 7–10, 13–21

Baryons, 73–74, 84
Beta decay, 32, 34–37, 71, 78–79
Beta rays (beta particles), 32, 34–37
Bosons, 73–75, 88, 90
Bubble chambers, 46, 52–54

Charge, 11–16, 25, 35, 47, 64, 68, 70, 78–79, 90–91
Charm, 67, 85
Color, 67–68, 85–87
Conservation laws, 78–80, 90
Cosmic radiation, 41–44, 47

Decay, 22–23, 27–38, 43, 46, 52, 57, 62–63, 79

Electricity, 10–12, 16, 25–26, 47, 90

Electromagnetic force, 15–16, 25, 36, 47–49, 75, 83, 88–90, 94–95
Electromagnetic radiation, 25–27
Electron volts, 46–47
Electrons, 10, 13–16, 18, 21–22, 27, 34–38, 40–41, 45–47, 52, 61, 68–71, 73–74, 78, 82, 87, 90–91
Elements, 7–10, 13, 16–17, 21–23, 27–34, 37, 41, 45, 62–63, 82
Energy, relation to matter, 18, 23–24, 35, 39, 61, 71, 77–78

Fermions, 73–75, 85
Fields, 88–89
Flavor, 67, 85–87
Fundamental particles, 82–87

Gamma decay, 32, 37–38
Gamma rays (gamma particles), 26, 32, 37–38, 69
Gravitational force, 4–5, 25, 36, 75, 83, 88–89, 95

Hadrons, 73–75, 84–86, 89
Half-life, 30–31, 62

99

Index

High-energy particle collisions, 39–46, 52–54, 57–59, 76, 78, 86, 93–94

Isotopes, 29, 37, 45

Leptons, 73–75, 86–87, 89, 91, 93
Light, 24–27, 72

Mass, 24–25, 35, 39, 61, 70–71, 77–78
Matter, 5–7, 18, 23–24, 26, 61, 71–72, 77, 82, 87
Mesons, 73–75, 84, 88
Muons, 73, 87, 93–94

Nuclear force, strong (nuclear binding force), 16–18, 21, 25, 28, 34, 36, 74, 83, 88–89, 95
Nuclear force, weak (see also weak interactions), 36, 75, 83, 88–90, 94–95
Nucleus, 14–18, 21–23, 28–29, 32–34, 39–41, 45, 62
Neutrinos, 34–36, 38, 73, 78, 87, 90–94
Neutrons, 15–17, 21–23, 27–30, 32–35, 37, 40–41, 47, 61, 73–74, 78, 82

Parity, 79–80, 90
Photons, 25–27, 35, 38, 65, 70, 73–75, 88–90
Protons, 10, 13-18, 21–23, 27–30, 32–34, 37–38, 40–41, 45–47, 52, 61, 70, 73–74, 78, 82

Quantum numbers, 64–68, 72–74, 79, 85
Quarks, 84–87

Radiation, 23–38, 41–44, 52, 54
Radioactivity, 27–38

Spark chambers, 46, 54–57
Spin, 65–66, 79
Strangeness, 66, 79
Subnuclear particles, definition of, 20, 38–40, 61–63, 82–83

Track photographs, 46, 52–59, 80, 92–94

Unified field theory, 83, 88–89, 95

W-particles, 75, 89, 91
Weak interactions, 36, 79, 90–95

PHOTO CREDITS

FRONTIS: Courtesy of the Stanford Linear Accelerator Center, Stanford University.

Page 8—Courtesy Hale Observatories.
Page 19—Courtesy Hale Observatories.
Page 48—Courtesy Fermi National Accelerator Laboratory.
Page 50—Courtesy Fermi National Accelerator Laboratory.
Page 51—Courtesy Stanford Linear Accelerator Center, Stanford University.
Page 53—Courtesy Fermi National Accelerator Laboratory.
Page 55—Courtesy Brookhaven National Laboratory.
Page 56—Courtesy Argonne National Laboratory.
Page 58—Courtesy Brookhaven National Laboratory.
Page 67—Photo courtesy Brookhaven National Laboratory. Superimposed artwork reprinted with permission from SCIENCE NEWS, the weekly news magazine of science, copyright © 1976 by Science Service, Inc.
Page 69—Courtesy Brookhaven National Laboratory.
Page 92—Courtesy Argonne National Laboratory.